# CONTENTS

Introduction

Agriculture
- wheat
- oilseed rape
- improved rough grazing

Woodlands
- removal of woodland
- new woodland
- young individual trees

Boundaries
- removal of boundaries
- new boundaries
- old hedgerows

Buildings & roads
- new housing estates
- new roads and tracks
- new agricultural buildings

Recreation
- golf courses
- caravan sites
- horses

Summary

# INTRODUCTION

The Institute of Terrestrial Ecology (ITE), as an independent research institute of the Natural Environment Research Council, has a national responsibility for the monitoring of changes in rural land use and landscapes, as an essential part of its role in assessing the implications of past, present and future policies on the terrestrial and freshwater environment. This responsibility is mirrored by the general interest expressed by groups representing all types of land users, government agencies, planning authorities and parliamentary committees. No issue concerned with land use is more keenly debated than that of landscape change, and, in particular, the long-term environmental effects of differing uses of land.

It was in seeking to confront these important issues that ITE has conducted 2 surveys of land use, one in 1977–78 and the other in 1984, ie a gap of 6 years, in order to determine the recent changes taking place in characteristic land uses, and hence in the resulting landscapes. The method involves sampling a representative series of sites throughout Great Britain, each site measuring 1km × 1km. Details of how these surveys were done, and discussion of sampling errors, are presented elsewhere. This short booklet is intended to present examples of some of the more important results.

The ITE surveys covered a wide range of land use and landscape features, but this presentation concentrates on 15 features which are of topical interest, 3 each from the broad categories of agriculture, woodlands, boundaries, the built environment and recreation. (Information on other features will be available in the more detailed descriptions of the results currently being prepared.)

Maps showing the general occurrence of each feature in 12 regions of Britain are included in this presentation. These regions are based on counties that have essentially similar land characteristics, although county boundaries are not generally decided by environmental (and hence land cover type) criteria. Cumbria, for example, includes nearly the full range of agricultural land cover types found in Britain. By contrast, Lincolnshire or Cambridgeshire tend to be more homogeneous in their land form and land use.

The information given in this document represents some objectively derived findings from the 2 ITE surveys. Judgemental interpretation has been kept to a minimum, although analysis of ecological effects of landscape change may be reported in future publications.

# AGRICULTURE

- wheat
- oilseed rape
- improved rough grazing

# AGRICULTURE

# WHEAT

The area of wheat was recorded at each sample site in 1978 and 1984 and, from measurements of these areas, national and regional estimates have been calculated. Winter and spring wheat crops were not differentiated.

The estimated total area of wheat in Great Britain in 1978 was 1 053 600 hectares and had increased to 1 748 500 hectares by 1984. This difference represents an increase of 66%.

Wheat was present in 24% of sites in 1978 but had spread to 33% by 1984. The greatest increases in the area of wheat were in East Anglia but percentage changes were most dramatic in south-west England (from 1·7% to 5·4% by area).

Analyses of changes in land use between survey dates show that 93% of land growing wheat in 1984 had been under annual crops or short-term grass leys in 1978. This finding confirms that the increase in the area of wheat has been at the expense of other crops (especially barley) rather than through the cultivation of new ground.

Comparison of areas under wheat in 1978 and 1984

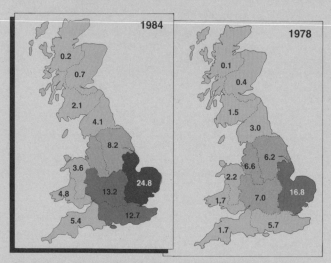

Comparison of land under wheat in 1978 and 1984, shown as % of the total land area in each of 12 regions

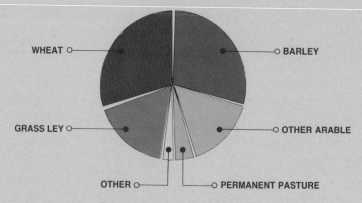

Previous use of land under wheat in 1984

## — area had increased by 66% and was spreading west and north

# OILSEED RAPE

Comparison of areas under oilseed rape in 1978 and 1984

Fields of rape were recorded at the sample sites in 1978 and 1984, and national and regional estimates were calculated.

The estimated total area of oilseed rape was 18 200 hectares in 1978 and had increased 10-fold by 1984 to 183 000 hectares. The 1984 total represents less than 1% of the land area of Great Britain. In 1984, the largest areas of rape were found in central and east England.

Of the land growing rape in 1984, 96% had previously been used to grow annual crops or short-term grass leys in 1978.

The period between surveys covers the rise in popularity of oilseed rape as a break-crop in agricultural rotations, but it has been suggested that rape production may have reached a peak.

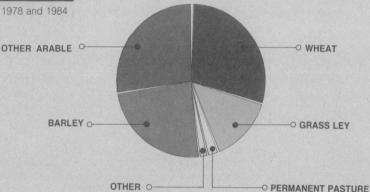

Previous use of land under oilseed rape in 1984

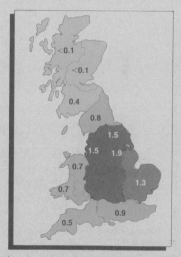

Land under oilseed rape in 1984, shown as % of the total land area in each of 12 regions

**— area had increased ten-fold but was still only 1% of Britain**

AGRICULTURE

# IMPROVED ROUGH GRAZING

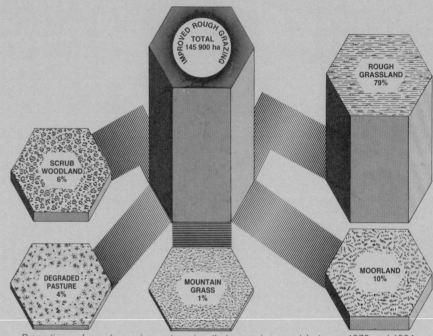

Proportions of rough grazing categories that were improved between 1978 and 1984

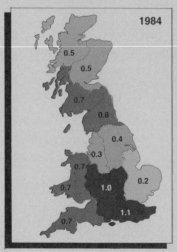

Land where rough grazing had been improved between 1978 and 1984, shown as % of the total land area in each of 12 regions

Land that was recorded as rough grazing in 1978 included upland grass, moorland and mountain vegetation. By 1984, some of this land had been agriculturally upgraded, often by drainage and re-seeding.

The estimated total area of improved rough grazing in 1984 was 146 000 hectares. This figure is equivalent to about 3% of the total rough grazing area in 1978.

The areas where improvement was most frequent included south and central England as well as north and mid-Wales. Drainage in other upland areas did not improve land out of the rough grazing category.

In half the cases where land had been improved, drainage had been carried out, but most of the wetland sites identified in the 1978 survey were left undrained.

— **about 3% of rough grazing had been upgraded**

# WOODLANDS

- removal of woodland
- new woodland
- young individual trees

# REMOVAL OF WOODLAND

The areas of broadleaf and coniferous woodland which had been removed or replaced between the survey dates were measured. Included in the resulting national and regional estimates are areas of broadleaf woodland that had been underplanted with conifers.

The total area of removed woodland was 70 800 hectares, of which 24 700 were broadleaf woodland and scrub, and 46 100 coniferous. In addition, 11 200 hectares of broadleaf woodland had been underplanted with conifers.

70 800 HA OF WOODLAND WERE REMOVED, 1978–84

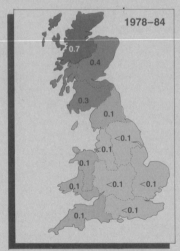

Land where broadleaf woodland and scrub was removed between 1978 and 1984, shown as % of the total land area in each of 12 regions

Land where conifers were felled between 1978 and 1984, shown as % of the total land area in each of 12 regions

Most of the lost broadleaf woodland was in south and central England, but nearly all of the felled coniferous forest was in Scotland.

The area of lost broadleaf woodland and scrub included 39% which had been replaced by, or underplanted with, conifers. Of the remainder, most had been converted to agricultural use. Nearly all of the clear-felled coniferous woods had been re-planted with second-rotation conifer crops.

Relative areas and subsequent land use of woodland removed between 1978 and 1984

## — about 6% of all broadleaf woodland had been lost

# NEW WOODLAND

|  | BROADLEAF | CONIFER |
|---|---|---|
| LINES | 8000 km | 1600 km |
| AREAS | 26000 ha | 177000 ha |

Lines and areas of new woodlands in 1984

Trees which had been planted between surveys were measured as lines or areas (see overleaf for numbers of individual trees). Excluded from these estimates are trees planted on golf courses or other formal recreation areas.

The total area of new broadleaf planting (including small areas of mixed broadleaf/conifer woodland) was 26000 hectares. The total area of new conifer plantation was 177000 hectares. In addition, trees were planted in lines totalling 9600 kilometres, of which 83% were broadleaf.

The planting of areas of broadleaf trees was most common in south and central England, while lines of trees were more frequent in East Anglia. More new coniferous plantations were found in Scotland.

The previous land use of new broadleaf areas was usually farmland, and often planting had taken place in field corners or at the edges of fields or roads. By contrast, coniferous plantings were on a large scale in upland grassland and moorland areas.

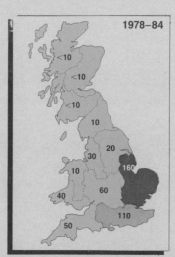

Lines of trees planted between 1978 and 1984, shown in metres per kilometre square in each of 12 regions

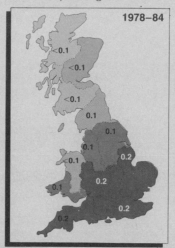

Land planted with broadleaf trees between 1978 and 1984, shown as % of the total land area in each of 12 regions

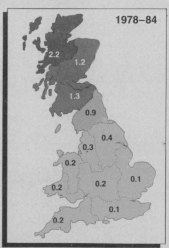

Land planted with conifers between 1978 and 1984, shown as % of the total land area in each of 12 regions

**— more broadleaf woodland was planted than had been removed**

# YOUNG INDIVIDUAL TREES

The numbers of individual young trees, estimated to be 5 years old or younger, were recorded at each sample site. Groups of 10 or more young trees were classified as new woodland (see previous page). Trees growing on golf courses and other recreation areas were excluded from these estimates. Young trees in hedgerows were differentiated from other isolated trees.

The estimated number of young hedgerow trees in Great Britain was 59 000. Other isolated young trees totalled 161 000, of which most were planted.

About half of the total number of young trees were found in East Anglia where small-scale tree planting schemes and hedgerow tagging are becoming common.

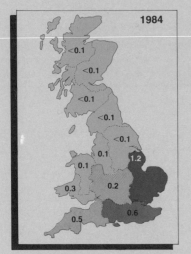

Young hedgerow trees in 1984, shown as the number per kilometre square in each of 12 regions

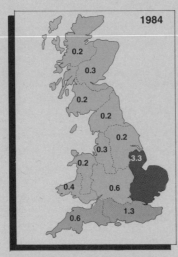

Young isolated broadleaf trees in 1984, shown as the number per kilometre square in each of 12 regions

The national average density of young trees was about one tree per kilometre square, with areas away from East Anglia having lower densities.

Survey reports suggest that many victims of Dutch elm disease were suckering and may provide a valuable future resource.

## — more were found in East Anglia than elsewhere

# BOUNDARIES

- removal of boundaries
- new boundaries
- old hedgerows

BOUNDARIES

# REMOVAL OF BOUNDARIES

The lengths of whole boundaries that had been removed between survey dates were measured. Replacement of one boundary type by another is not considered here.

In the 6-year period between surveys, 28 000 kilometres of hedgerow had been removed; 12 000 kilometres of wire fence and 1 400 kilometres of wall had also been removed and not replaced.

Hedgerows were lost from most regions in England and Wales, with rates of removal in East Anglia being no greater than elsewhere.

Wire fences had been removed from most regions and the relatively short lengths of wall were lost from north England and the Scottish regions.

Approximately 8 times as much hedgerow had been removed as had been planted.

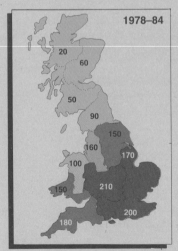

Lengths of hedgerow removed between 1978 and 1984, shown in metres per kilometre square in each of 12 regions

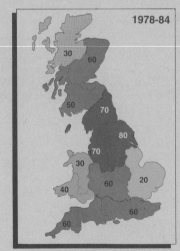

Lengths of wire fence removed between 1978 and 1984, shown in metres per kilometre square in each of 12 regions

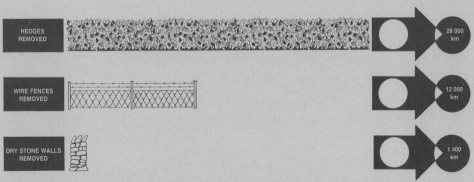

Comparison of lengths of boundary types which have been removed between 1978 and 1984

## — more than half the boundaries removed were hedges

# NEW BOUNDARIES

The lengths of new boundaries, established between survey dates, were measured. National and regional estimates do not include replacement of one boundary type by another.

Wire fences were the most common new boundary type totalling 48 400 kilometres. New hedges amounted to 3 500 kilometres and there were 13 700 kilometres of new wooden fences. There were no new walls.

New wire fences were found most commonly in central, south and south-west England. The new hedges were most frequent in south and south-west England, while new wooden fences were characteristic of East Anglia and south-east England.

The length of completely new boundary was more than twice the length of boundary which had been removed; 74% was wire fence, often associated with agricultural improvement or forestry.

Comparison of lengths of new boundary types established between 1978 and 1984

- 48 400 km — NEW WIRE FENCES
- 13 700 km — NEW WOODEN FENCES
- 3 500 km — NEW HEDGES

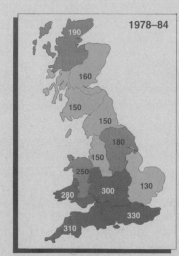

Lengths of wire fence erected between 1978 and 1984, shown in metres per kilometre square in each of 12 regions

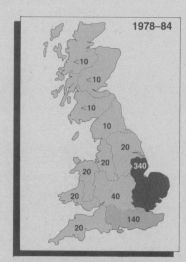

Lengths of wooden fence erected between 1978 and 1984, shown in metres per kilometre square in each of 12 regions

**— most were wire fences, but 20% were wooden fences**

# OLD HEDGEROWS

Total length of derelict hedgerow in 1984

Although difficult to define, derelict and relic hedgerows were recorded in the sample sites. Derelict hedgerows resembled the appearance of hedges but were no longer of use as stockproof boundaries. Relic hedgerows were often no more than lines of trees or shrubs. Both are interesting features of the landscape which are likely to change in the future.

In 1984, the length of derelict hedgerow in Great Britain was 58 000 kilometres with a further 46 600 kilometres of relic hedgerow.

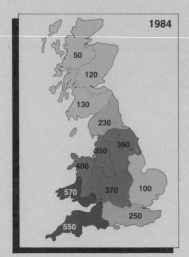

Lengths of derelict hedgerow in 1984, shown in metres per kilometre square in each of 12 regions

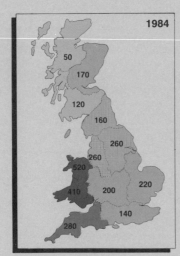

Lengths of relic hedgerow in 1984, shown in metres per kilometre square in each of 12 regions

These features were especially characteristic of south-west England and Wales, but these figures may reflect the greater overall lengths of hedgerow in these regions.

The future changes in these features are uncertain. They may be removed (as a result of agricultural intensification or as a fuel-wood source) or they may grow into lines of trees, being often uneconomic to re-lay.

Total length of relic hedgerow in 1984

**— were characteristic of western England & Wales**

# BUILDINGS & ROADS

- new housing estates
- new roads and tracks
- new agricultural buildings

# NEW HOUSING ESTATES

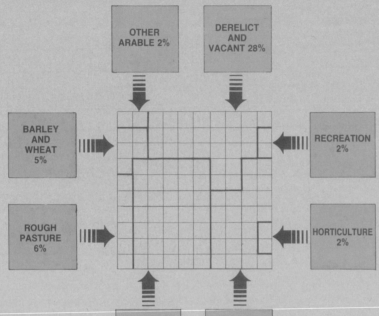

Previous use of land on which houses were built between 1978 and 1984

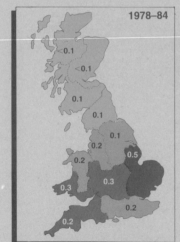

Areas of new housing estates built between 1978 and 1984, shown as % of the total land area in each of 12 regions

Although the surveys were designed to sample the rural environment, many features of the urban fringe were recorded. Housing estates built in the period between surveys were measured by area, and national and regional estimates were made.

The total area of new housing estates was 39 000 hectares.

Most building had taken place in East Anglia, with fewer estates in north England and Scotland.

Building had most frequently taken place on land which had previously been vacant, derelict or under pasture.

**— were usually built on derelict land or agricultural grassland**

# NEW ROADS & TRACKS

Tarmac roads and constructed tracks which had been built between survey dates were measured. Only those in rural situations (but excluding forestry plantations) have been considered here. National and regional estimates have been obtained.

The total length of new tarmac road was 4 300 kilometres and the length of new tracks was 5 000 kilometres.

The region with most road building was central England, while most of the newly constructed tracks were found in south-west England, Wales and Scotland. Many of the tracks were found in lowland areas and were often associated with agricultural intensification.

Total length of new tracks constructed between 1978 and 1984

Total length of new roads built between 1978 and 1984

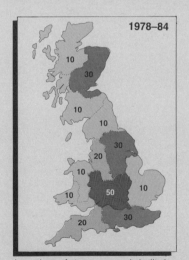

Lengths of tarmac road built between 1978 and 1984, shown in metres per kilometre square in each of 12 regions

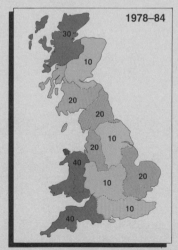

Lengths of track constructed between 1978 and 1984, shown in metres per kilometre square in each of 12 regions

— lengths of tarmac road were matched by new farm & upland tracks

# NEW AGRICULTURAL BUILDINGS

Total number of new agricultural buildings in 1984

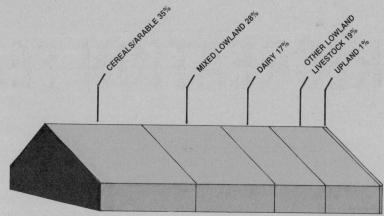

New farm buildings according to farm type

The areas of farm buildings were recorded in 1978, and new structures which extended these areas, or formed new sites, were recorded in 1984.

The number of new farm buildings was 105 500.

There were no strong regional trends, except that southern regions had greater numbers of new farm buildings than northern, and especially Scottish, regions.

If there are about 220 000 farms in Britain, then, on average, nearly 50% of them had one new building erected in the 6-year period between surveys.

New farm buildings were most frequently found on cereal/arable farms and only 1% were associated with upland farming systems.

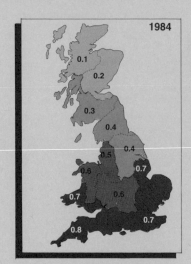

New agricultural buildings in 1984, shown as the number per kilometre square in each of 12 regions

## — on average, 50% of farms had new buildings

# RECREATION FEATURES

- golf courses
- caravan sites
- horses

RECREATION

# GOLF COURSES

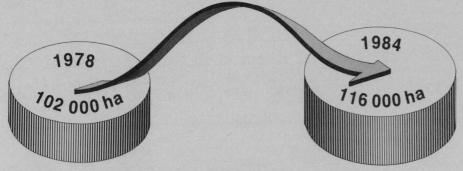

Comparison in areas of golf courses in 1978 and 1984

The area of golf courses was measured in the sample sites in 1978 and in 1984, and national and regional estimates were made for the change in area.

The total area of golf courses in 1978 was 102 000 hectares and had increased by 14% to 116 000 hectares by 1984; 98% of the increased area was due to the expansion of existing sites.

Golf courses were found most commonly in English regions, especially in the central region.

Where golf courses had increased in area, it was usually at the expense of agricultural, often arable, land.

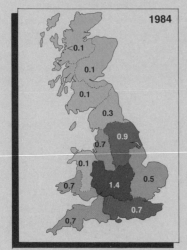

Areas of golf courses in 1984, shown as % of the total land area in each of 12 regions

| | |
|---|---|
| ARABLE | 82% |
| PERMANENT PASTURE | 4% |
| ROUGH GAZING | 14% |

Previous use of land which had been converted to golf courses between 1978 and 1984

— increase in area had resulted from development of existing sites

# CARAVAN SITES

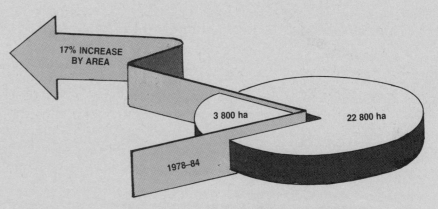

The area by which caravan sites had increased between 1978 and 1984

This category included formal caravan sites and parks. The areas of each were measured in the sample sites in 1978 and 1984, and combined estimates have been made for the regions and for Great Britain as a whole.

The total area of formal caravan sites in Great Britain was 22 800 hectares in 1978 and 26 600 hectares in 1984, representing a 17% increase.

Formal sites were present in all regions but were less frequent in East Anglia and Scotland (where informal or single caravans were more usual). Caravan sites were especially associated with coastal locations.

The number of individual caravan sites had increased by 11% between 1978 and 1984.

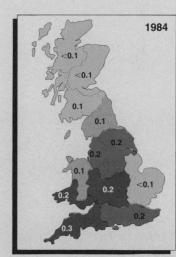

Areas of formal caravan sites in 1984, shown as % of the total land area in each of 12 regions

**— new sites had contributed to increase in area**

RECREATION

# HORSES

The number of horses in each of the sample sites was recorded in 1984, and national and regional totals were calculated. Only horses which were kept outside at the time of survey (summer) were recorded and no distinction between types of horse was made.

The total number of horses in Great Britain in 1984 was 558 400.

Total number of horses in 1984

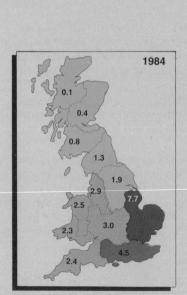

Horses in 1984, shown as the number per kilometre square in each of 12 regions

The area of different types of land on which horses were found

Horses were usually recorded on their own in fields, but in 28% of occurrences they were found with agricultural stock.

The most frequent grazing land on which horses were recorded was described as permanent pasture (62%), with rough grazing (33%) featuring in upland areas. Short-term grass leys accounted for only 5% of the land on which horses were found.

Horses were recorded in 31% of the sample sites but most were found in East Anglia and south England. The high numbers of horses in East Anglia were usually found in urban fringe situations but also on stud-farms.

If a horse uses about 2 hectares of grass per year, then these figures suggest that horses in Great Britain need about one million hectares of land, or about 15% of the lowland grass.

— may be using up to 15% of lowland grass in Britain

# SUMMARY

ITE carried out field surveys which sampled Great Britain in 1978 and 1984.

Both surveys yielded information on land cover types, land use and landscape features, with the second survey being particularly detailed. Changes in the extent and distribution of certain categories have been calculated. For the purposes of the report, some examples of the types of information, and the results associated with their assessment, are presented as a fore-runner to a fuller report to be published in due course.

The changes in wheat, oilseed rape and improvement of rough grazing were analysed as examples of how agricultural land cover types have been surveyed. Although there may have been variations in the intervening years, the area of land under wheat had increased by 66% between survey dates. Oilseed rape had increased 10-fold and, like wheat, occupied land that was previously under cereals, other arable crops or short-term grass. The analysis of crop changes, field by field, showed that only a small percentage (about 3%) of the new cereal land had come from permanent pasture or poorer grazing land.

When considering woodland changes, it was shown that the area of newly planted broadleaf trees approximately equalled the area of broadleaf woodland and scrub removed. However, the new plantings differed from the ancient broadleaf woods, and much of the planting had taken place in already depleted East Anglia. The removal of woodland continued in neighbouring areas such as central, south and south-east England. In addition, a relatively large percentage of broadleaf woodland had been underplanted with conifers.

Most of the removal of conifer forest was associated with re-planting of conifers in Scotland, especially in the highlands and islands. Some new coniferous plantations (about 9%) had been established where broadleaf woodland had been removed.

Many young trees had been planted, especially in East Anglia, but the number of young hedgerow and isolated trees was small (one tree per kilometre square) when considered on a national scale.

The removal of boundaries was recorded. Minimum figures for the removal of whole hedgerows (excluding those replaced by another boundary type) indicated that 28 000 kilometres had been removed. Much of this removal had taken place in areas where cereal crops had increased, such as central, south and south-east England. By contrast, new boundaries were mainly wire fences (48 400 kilometres), although there were some new hedges and wooden fences, the latter usually being associated with horses. There were about 8 times as many hedges removed as had been planted.

New built features included housing estates (39 300 hectares), agricultural buildings (105 000) and roads and tracks (4 300 and 5 000 kilometres respectively). Much of the new housing was found in East Anglia and other regions within commuting distance of London. Agricultural buildings were found most frequently in the lowland regions, as were roads and tracks. In the Scottish islands, many new tracks were built.

Recreation topics represented in the report were golf courses, caravan parks and horses. The areas of golf courses and caravan parks had both increased by about 15%, development of the former tending to be through enlargement of existing sites.

Interest in the increasing popularity of horse-ownership, especially in the urban fringe, led to the estimation of the number of horses, which exceeded half a million.

Overall, the major conclusion from the results of the ITE surveys was that the loss of landscape features was slowing down in East Anglia, but was probably increasing elsewhere in southern Britain, in association with the spread of cereals and the intensification of agriculture. Meanwhile, modification of the landscape was being carried out in East Anglia through the planting and encouragement of young trees.

The 2 ITE surveys have allowed estimates to be made of changes taking place in the British landscape. The information obtained from these surveys forms a baseline for future monitoring of the rural environment, as well as allowing study of the ecological effects of such change, and the prediction of future trends.